ANIMAL COVERINGS

FEATHERS

by Eric Geron

Children's Press®
An imprint of Scholastic Inc.

Library of Congress Cataloging-in-Publication Data

Names: Geron, Eric, author.
Title: Feathers / by Eric Geron.
Description: First edition. | New York : Children's Press, an imprint of Scholastic Inc., 2024. | Series: Learn about: animal coverings | Includes index. | Audience: Ages 5–7 | Audience: Grades K–1 | Summary: "Let's learn all about the different types of animal coverings! Animals have different body coverings for different reasons. Some animals use their coverings to keep warm or stay cool, others use them for protection, and can either stand out or blend in. Some animals even use their coverings to move! This vibrant new set of LEARN ABOUT books gives readers a close-up look at five different animal coverings, from fur and feathers to skin, scales, and shells. Each book is packed with photographs and fun facts that explore how each covering suits the habitat, diet, survival, and life cycle of various animals in the natural world. Which animals have feathers? Birds! Do you know why birds need feathers to survive? With amazing photos and lively text, this book explains how feathers help birds eat, fly, swim, stay warm, blend in, and more! Get ready to learn all about feathers!"—Provided by publisher.
Identifiers: LCCN 2023000187 (print) | LCCN 2023000188 (ebook) | ISBN 9781338897999 (library binding) | ISBN 9781338898002 (paperback) | ISBN 9781338898019 (ebook)
Subjects: LCSH: Feathers—Juvenile literature. | Birds—Adaptation—Juvenile literature. | Body covering (Anatomy)—Juvenile literature. | BISAC: JUVENILE NONFICTION / Animals / General | JUVENILE NONFICTION / Science & Nature / General (see also headings under Animals or Technology)
Classification: LCC QL697.4 .G47 2024 (print) | LCC QL697.4 (ebook) | DDC 598.147—dc23/eng/20230110
LC record available at https://lccn.loc.gov/2023000187
LC ebook record available at https://lccn.loc.gov/2023000188

A special thank-you to the team at the Cincinnati Zoo & Botanical Garden for their expert consultation.

Copyright © 2024 by Scholastic Inc.

All rights reserved. Published by Children's Press, an imprint of Scholastic Inc., *Publishers since 1920.* SCHOLASTIC, CHILDREN'S PRESS, and associated logos are trademarks and/or registered trademarks of Scholastic Inc.

The publisher does not have any control over and does not assume any responsibility for author or third-party websites or their content.

No part of this publication may be reproduced, stored in a retrieval system, or transmitted in any form or by any means, electronic, mechanical, photocopying, recording, or otherwise, without written permission of the publisher. For information regarding permission, write to Scholastic Inc., Attention: Permissions Department, 557 Broadway, New York, NY 10012.

10 9 8 7 6 5 4 3 2 1   24 25 26 27 28
Printed in China   62

First edition, 2024
Book design by Kay Petronio

Photos ©: 8: The Natural History Museum, London/Science Source; 9 all feathers: Andrew Leach/Cornell Lab of Ornithology; 12 bottom: Carlyn Iverson/Science Source; 17 inset: James Marvin Phelps/500px/Getty Images; 22 inset: Wirestock/Getty Images; 29 center right: Stefan Christmann/Nature Picture Library; 29 bottom left: Steve Daggar Photography/Getty Images; 29 bottom right: Kativ/Getty Images.

All other photos © Shutterstock.

CONTENTS

Introduction: Birds of a Feather 4

Chapter 1: Light as a Feather 8

Chapter 2: Feathers in Motion 12

Chapter 3: Weather Feathers 18

Chapter 4: Hiding 22

Chapter 5: What Else Can Feathers Do? 26

Conclusion: Feathers Matter 30

Glossary 31

Index/About the Author 32

INTRODUCTION
Birds of a Feather

Animal bodies can have different coverings. Some are covered with skin or fur. Others are covered with scales or shells. This book is all about a special covering: feathers! Feathers can be long and wide, or short and thin.

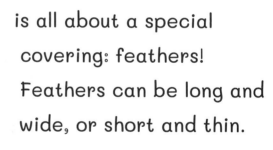

Close-up of peacock tail feathers

PEACOCK

Feathers can also make animals look colorful. Discover which animals have them, what feathers are made of, and the amazing things they can do.

Which Animals Have Feathers?

Birds have feathers! Birds are **warm-blooded** animals with beaks. They can lay eggs with hard shells. All birds have wings, and most birds can fly. Most birds are **omnivores**.

A group of birds is called a *flock*.

Their diet mainly consists of insects, fruits, nuts, and seeds. Some birds eat meat. Birds live all over the world.

Feathers are made from **keratin**, the same material that makes up human hair and nails.

CHAPTER 1
Light as a Feather

Feathers are really useful for birds. Just like there are different types of birds, there are also different types of feathers. Each type of feather plays an important role. Some feathers protect against wind and weather.

TYPES OF FEATHERS

Wing · Tail · Contour · Down · Bristle

CHICKEN

Some feathers **steer** birds through the air. Some feathers keep birds dry in the water or warm in the cold. Some feathers protect a bird's face or hide them from danger. Some feathers let a bird show off!

Out of the 10,000 **species** of birds in the world, the chicken is the most common.

HAWK

- Wing Feathers
- Bristle Feathers
- Contour Feathers
- Tail Feathers

This is a hawk. It uses its wing feathers to lift off and fly. Wing feathers are long, tough, and protect against wind.

Most birds have 12 tail feathers.

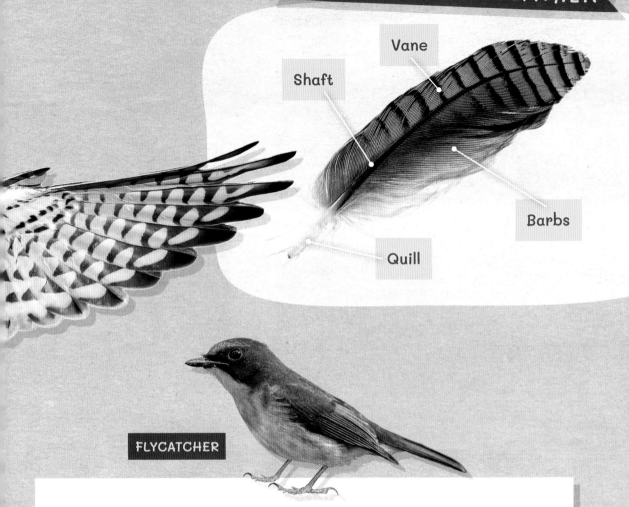

PARTS OF A WING FEATHER

- Vane
- Shaft
- Barbs
- Quill

FLYCATCHER

Tail feathers help birds with balance. They also help birds to steer and slow down. **Contour** feathers are strong and flexible. They form a smooth body shape so that a bird can fly with ease. Bristle feathers protect a bird's eyes and face.

CHAPTER 2

Feathers in Motion

Many birds are great fliers. When they are in the air, pointy wings are good for darting. Long wings are good for gliding.

Bird bones are hollow because it's easier to fly with less weight!

BIRD BONE

Some birds cannot fly at all, like penguins, kiwis, and ostriches!

KIWI

EAGLE

The bigger the bird, the slower it needs to beat its wings. Small birds beat their wings very fast, which requires a lot more energy.

Walking and Running

Almost all birds can walk. A lot of birds can hop, waddle, or sometimes even run. Birds mostly move across the ground when they are looking for food to eat. Their feathers help give them support when walking.

GEESE

OSTRICH

Close-up of ostrich feathers

When certain birds like ostriches run, they also use their wings for balance.

Swimming and Floating

Many birds are good swimmers, like ducks and penguins. Some birds press their wings to the sides of their bodies and dive into the water.

Close-up of Emperor penguin feathers

Penguins use the soft feathers on their stomachs to slide across ice and snow!

ADÉLIE PENGUINS

MANDARIN DUCK

Many birds can float. They use their feathers to trap air and float on the water's surface like balloons.

CHAPTER 3
Weather Feathers

Birds use their feathers to help control their body temperature by cooling off or warming up. One way a bird can cool down on a hot day is by going for a dip in the water.

RAINBOW LORIKEETS

SPARROW

Birds constantly clean and comb their feathers with their beaks.

FLAMINGO

Birds can cool down even more by fluffing up their feathers while drying off. When a bird spreads its wings, a breeze can run through the bird's feathers and cool off its warm skin.

Warming Up

Birds also use their feathers to trap heat close to their bodies. This allows them to stay warm in the cold. **Down** feathers are soft and fluffy. They are used to trap warm air. They keep birds warm even in cold water.

ROBIN

SWAN

Some birds can tuck their head, feet, and legs into their feathers for warmth. Others, like the snowy owl, have feathers on their legs and feet for extra warmth.

SNOWY OWL

Contour feathers also help to protect birds against the chilly wind and weather.

CHAPTER 4
Hiding

Feathers can sometimes help birds hide from hungry **predators**. The ability to blend into their surroundings is called **camouflage**. Camouflage helps protect birds from being seen by threats. Nightjars disappear into their surroundings since their feathers look like dead leaves.

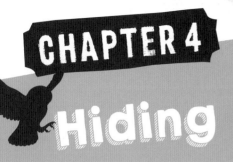

NIGHTJAR

When birds are attacked by predators, they can lose some of their tail feathers to help get away.

Some owls can blend into the bark of trees. Willow ptarmigans (TAR-mi-guhnz) look like part of a snowy white landscape.

EASTERN SCREECH OWL

WILLOW PTARMIGAN

Blending In

Camouflage is not only used to hide birds from dangerous predators. It can also be helpful to hide birds from their own **prey**. A bird's feathers make it easy to blend in and sneak up on their prey without being seen and scaring it off.

BARN OWL

Some birds have feathers that surround their faces in a disc shape. This shape collects sounds to help them hear better while hunting.

SANDGROUSE

Birds that hunt at night, like the owl, can keep quiet by using their feathery wings. Their feathers muffle any sounds they may make.

CHAPTER 5
What Else Can Feathers Do?

Birds use the colors and patterns of their feathers for more than just blending in. They also use them to stand out!

COCKATOO

Some birds that display their feathers to impress other birds are peacocks, blue crowned pigeons, and cockatoos.

VICTORIA CROWNED PIGEON

Showing off their colorful feathers is a way for many birds to attract a **mate**. It can also be a signal birds use to warn predators to stay away.

New Feathers

All birds lose old feathers and replace them with new ones. This is called **molting**. Birds can molt one to two times per year, depending on the type of bird. Many birds molt all their feathers at once. Others only molt some of their feathers. Molting gets rid of any damaged feathers.

Some birds use feathers to make their nests nice and cozy.

BIRD'S NEST

CARDINAL

Blue Jay

Emperor Penguin

New feathers grow and allow birds to keep flying, staying dry, or putting on a display.

Rainbow Lorikeet

CONCLUSION

AMERICAN KESTREL

Feathers Matter

Now you know all about feathers! They can be strong and flexible, and some can be soft and fluffy. They come in all shapes, sizes, and colors! Feathers can help birds with flight, warmth, and keeping dry. They also help birds hide from view or show off to other birds. Next time you see a bird flying through the sky, remember how its feathers make it possible.

> The bird with the most feathers is the whistling swan, with over 25,000 feathers!

WHISTLING SWAN

GLOSSARY

camouflage (KAM-uh-flahzh) a disguise or natural coloring that allows animals to hide by making them look like their surroundings

contour (KAHN-toor) the outline of an animal or object

down (doun) the soft feathers of a bird

flock (flahk) a group of animals of one kind that live, travel, or feed together, such as birds

keratin (KEHR-uh-tin) a fiber-like protein that forms the structure of feathers, hair, and nails

mate the male or female partner of a pair of animals

molt (mohlt) to lose old fur, feathers, shell, or skin so that a new layer can grow

omnivore (AHM-nuh-vor) an animal that eats both plants and meat

predator (PRED-uh-tur) an animal that lives by hunting other animals for food

prey (pray) an animal that is hunted by another animal for food

PARAKEET

species (SPEE-sheez) one of the groups into which animals and plants are divided

steer to guide or to direct

warm-blooded (WORM bluhd-id) having a body temperature that does not change, even if the temperature of the surroundings is very hot or very cold

31

INDEX

B
balancing, 11, 14–15
birds
  about, 6–7, 9
  flightless, 13
blending in, 22–25
bones, 12
bristle feathers, 9, 10–11, 24

C
camouflage, 22–25
contour feathers, 9, 10–11, 21
cooling off, 18–19

D
darting, 12
down feathers, 9, 20

F
feathers. See also bristle feathers; contour feathers; down feathers; tail feathers; wing feathers
  about, 4–5, 7, 30
  cleaning and combing, 18
  losing and replacing, 22, 28–29
  types of, 9–11
flightless birds, 13
floating, 17
flying, 12–13

G
gliding, 12–13

H
hiding, 22–25

L
lifting off, 10

P
protection, 10–11, 20–23

S
showing off and standing out, 5, 26–27
sliding, 16
steering and slowing down, 11
swimming and diving, 16

T
tail feathers, 9, 10–11
temperature control, 18–21

W
walking and running, 14–15
warming up, 20–21
wing feathers, 9, 10–11
wings, 12–13

ABOUT THE AUTHOR

Eric Geron is the author of many books. He lives in New York City with his tiny dog, who is as light as a feather.